MODERN VER... L

by
Michael and Victoria Roberts

Illustrated by
Sara Roadnight
Photographs: *Michael Roberts*

Cover Photograph: *Richard Roadnight*
A Mink near the Domestic Fowl Trust having killed a Buff Sussex cockerel.

© Michael and Victoria Roberts 1992
ISBN 0 947870 10 5

Published by the Domestic Fowl Trust,
Honeybourne, Evesham. Worcestershire
Telephone: (0386) 833083

Printed by Philip Bennett Lithographic Printers Ltd.
Stratford-upon-Avon.

Contents

		page
Introduction		4
What is Vermin?		5
The Law		6
Prevention		7
Methods		9
1. Snare	fox	9
2. Fenn trap	rats, grey squirrel, stoats, weasels, mink, feral ferrets	14
3. Cage trap	mink, grey squirrel	27
4. Mouse trap	mice	30
5. Mole trap	moles	32
6. Bird cage traps	rooks, carrion crows, magpies, jays, jackdaws, feral pigeons, collared doves, sparrows, moorhens, starlings	34
7. Poison (bait boxes)	rats, mice, grey squirrel	38
8. Ultrasonic devices	rats, mice, feral pigeons, starlings, moles	42
Identifying vermin from carcases		43
Useful by-products of vermin		44
Brown owls, badgers, hedgehogs, feral cats		46
Suppliers and useful addresses		48
Index		50

Introduction

We have always had a great interest in control of vermin and trapping since young, as nearly all school holidays were spent with gamekeepers. But we have been spurred to write this book by a plague of rats which arrived at the Domestic Fowl Trust one winter and by what, we wondered, other people would have done. We know of those who also suffered with a plague of rats and paid £200 to have a Council "rodent operative" to deal with the problem. Apart from rats there are times when you may be bedevilled with foxes, mink, stoats etc. and it is not easy to find someone with experience or specialised literature on how to deal with vermin. Too many books and leaflets pussyfoot around the problem and do not give you the vital tips on how to make a 100% kill, humanely.

You will notice that we have not mentioned very often the use of the gun. As a young trapper, the use of the gun from the end of March to the beginning of September was forbidden for fear of disturbing the game. Traps are working for you on a 24 hour basis, and not everyone has the desire, skill or licence to use a gun effectively.

It must be stressed that all the methods described for taking vermin are legal; but every care must always be made to ensure that there is no cruelty or suffering in applying these methods.

Successful control of vermin is like being a detective, looking for signs, tracks, food husks etc.; minute things that tell a story. This is gained by experience, patience, time, keen eyesight and observation.

This book is not about total eradication of predatory birds and animals. It is about slightly tipping the balance of nature in favour of mankind when vermin, certain birds and animals, become a plague or nuisance.

Michael and Victoria Roberts
Honeybourne, 1992

WHAT IS VERMIN?

The word vermin derives from the Latin *vermis*, a worm, and its dictionary definition is "small animals collectively, especially insects and rodents, that are troublesome to man, domestic animals etc." Modern usage extends to the avian varieties as well. Vermin can be very destructive, not only contaminating food, destroying birds and animals, but costly, too.

Winged vermin: Carrion crows, rooks, magpies, jackdaws, jays, house sparrows.
Ground vermin: Rats, mice, weasels, stoats, mink, fox, grey squirrel, feral ferrets.
Underground vermin: moles.
You will notice that we have not included the greater black-backed gull, hooded crow (although there is a similarity to the carrion crow) or coypu, none of which we have had any experience of controlling, but are included in Schedule II, Part 2 of the Wildlife and Countryside Act, 1981.

Mention will be made later of brown owls, badgers, hedgehogs, polecat, feral cats, moorhens and coots, so that you realise the potential threat from them and the laws that protect them.

We have not included rabbits and wood pigeon. These are pests, but there are gamekeeping publications available which deal more than adequately with these.

Before you embark on the control of vermin it is vital to know the law and its complexities.

Fox Tracks (see page 9)

THE LAW

The control of vermin and pests is covered by the following Acts of Parliament:

The Wildlife and Countryside Act 1981) Shooting, trapping or control
The Badgers Act 1973)
The Pest Act 1954   Trapping
The Agriculture Act 1947) Poison gas
The Agriculture Act (Scotland) 1948)
The Protection of Mammals Act 1911   )
The Protection of Animals (Scotland) Act 1912   ) Poisons
Animals (Cruel Poison) Act 1962   )
Animals (Miscellaneous Provisions) Act 1972   )

These Acts are complex, so here is basically what you may or may not do within the framework of the above Acts:

* All birds of prey and owls are protected.
* All winged vermin including crows, rooks etc. may be shot during the day or taken only in cage traps. They may not be poisoned, snared or spring trapped. Under special circumstances these winged vermin can assume plague proportions and the Ministry of Agriculture called in.
* Ground vermin i.e. rats, mice, weasels, stoats, fox, mink, grey squirrel, feral ferrets may be shot or trapped.
* Underground vermin - moles - may be shot or trapped.
* Poisons: certain poisons may be used against rats, grey squirrel and moles
* Badgers may be taken only under the supervision of the Ministry of Agriculture.
* Bird lime and gin traps are illegal and there are heavy penalties if found using either of these.
* Hedgehogs, polecats, feral cats may not be poison gassed, trapped or snared.
  These may be shot if there is conclusive evidence that they are causing trouble and hardship.
* Moorhens and coots may only be shot or cage trapped betweeen 1st September and 31st January.
* The otter is fully protected.

The legal information given in this book is believed to be correct at the time of going to press, but changes in the law do occur from time to time. If in doubt check with your local MAFF office, the CLA, the Game Conservancy, BASC or the BFSS.

In countries outside the United Kingdom readers are strongly advised to check with the legal department of their governments concerning trapping poisoning, or shooting any of the species mentioned in this book.

PREVENTION

As with all problems there is a cause and therefore measures which can be taken to prevent it. Where there is a concentration of anything, be it poultry, pheasants, pigs or parrakeets, this will be a draw for vermin. Basically vermin are after foodstuffs, the birds or animals themselves and/or the eggs or progeny.

Most modern foodstuffs arrive in bulk or paper bags. Proper storage facilities are important, either in metal bins, tanks or dustbins. The way your feed requirements are ordered i.e. not having too much feed lying around at any one time, and proper feed house or storage area add up to help minimise the problems from rats, mice and sparrows.

Methods of feeding are vital. It is crazy, having stored the feed correctly, then to throw it away by poor feeding practices such as leaving uneaten feed in a place accessible by vermin. This is really another aspect of good management.

Also important is the housing or fencing of your livestock, particularly if vulnerable to fox, mink, stoats, rats etc. Too many people buy expensive birds, thinking that as they can fly they are automatically safe, only to have them killed shortly afterwards by a fox. The following illustrations of fox-proof fencing are proven to be effective.

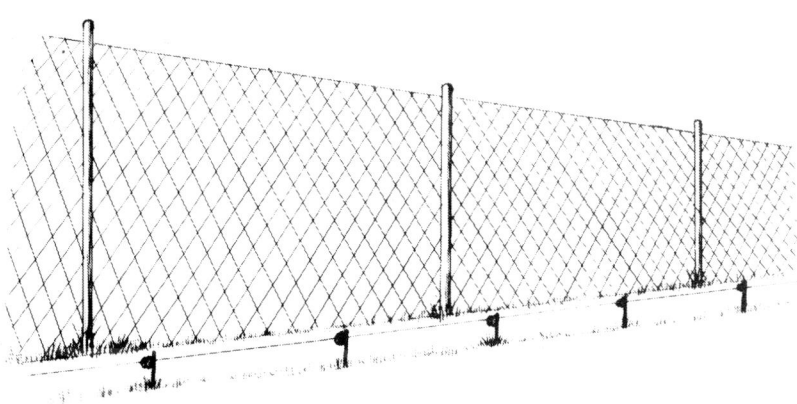

Fool proof anti-fox fence

Three styles of fox-proof fencing.

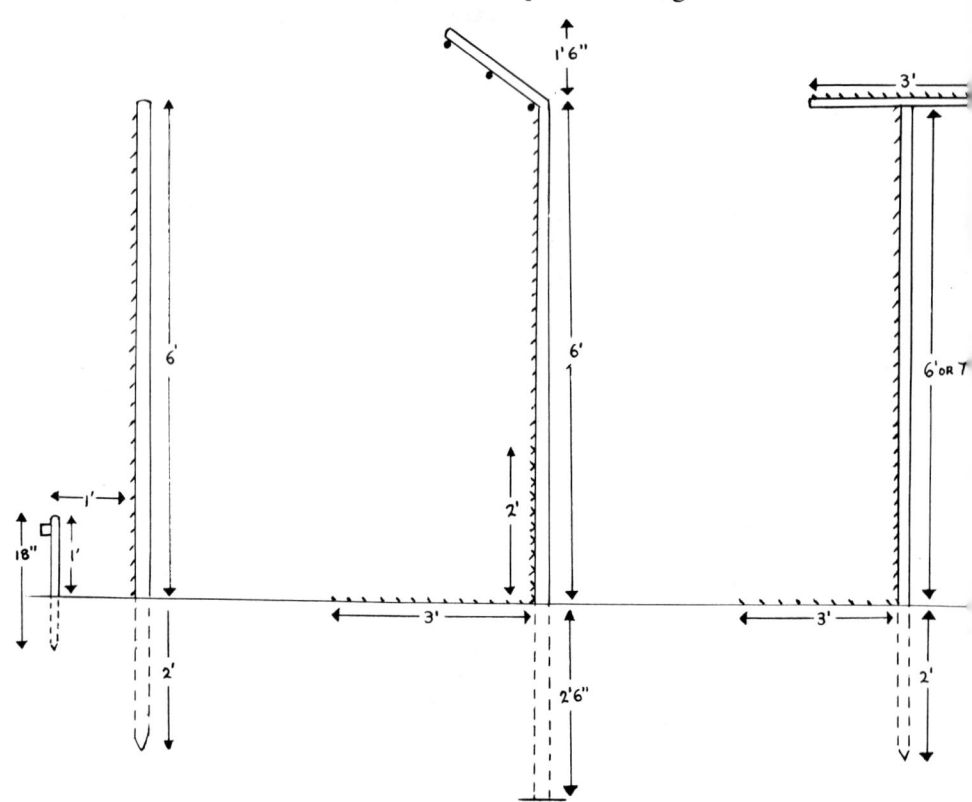

1. 8' wooden posts with 6'2" mesh wire netting and electric fence.

2. 2" x 2" angle iron with three strands of barbed wire at the top, then chain link or 2" wire mesh with 5' x 1" wire mesh on ground and up the side.

3. 8' or 9' wooden posts with 3' x 1" mesh wire on top, then pig or sheep wire then 5' or 6' x 1" mesh on ground and up the side.

Fence posts will always stay firm in the ground if they are put in in the winter when the ground is fairly soft.

GLOVES AND BOOTS

Before you start successful trapping it is essential to realise that human scent, soaps etc. are very strong. The animal world is mostly dependant on its specialised sense of smell for survival. Therefore you must have some strong, preferably rubber, gloves, and old boots. This is particularly essential for the amateur trapper or someone who hasn't got his or her hands and feet in the soil all day. Chemicals can also linger on boots, so it is sensible to have a set of boots and gloves for trapping only.

METHODS

1. **Snare**: quarry - fox

Foxes are the bane of all poultry, pheasant and duck keepers and in a single day or night will ruin a year or a lifetime's work. Their main active times of the year are in January and February and after the corn is cut. That is not to say that they are inactive at other times. In January and February they have young cubs to feed with not much young wild livestock to rely on, and when the corn is cut the cubs become more adventurous with enticing smells wafting over stubble.

Foxes can be shot in fox drives through woods, thickets etc. during the winter months, but care should be taken not to upset the local hunt. The shooting of foxes at night with a lamp is highly productive providing you know the lie of the land and there is no danger to livestock. Foxes are naturally curious and will come hundreds of yards to the squeal of a rabbit (caller) on a wet and windy night. The use of heavy shot such as 'BB' or 'SSG' or a .22 rifle is essential for a clean kill.

For snaring, the only snares available are the free running ones with swivels. The self locking type are now illegal. You will require a pricker (tealer) best made of hazel, but willow will do, together with a strong metal peg. A piece of angle iron 2' long with a ½" hole near the top is ideal.

Next comes the detective work. Where is the best place to set the snare? All foxes have set tracks or runs which they use night after night and in doing so travel many miles in a single night. You can tell when a fox has been around by the pungent scent it leaves which seems to hang in the air like pockets of mist. A fox run is a faint track, best seen through long grass, leading to a point through a fence, a gap in wire, under an old hurdle or gate. Look for the distinctive red-brown hair where he has passed through. It is here that you set the snare, where the fox has been scrabbling through or under an obstacle. When approaching the snaring point always do so down the track of the fox.

Great care should be taken to distinguish the track of a badger from a fox. In general a badger's track is more defined, well trodden, and brambles or roots in the way are chewed back. Although a fox may occasionally use a badger track, remember it is illegal to take badgers, so if in doubt, do not use the snare there.

Fox with winter coat

Set your snare as per the illustration, roughly a palm's width (4″) off the ground.

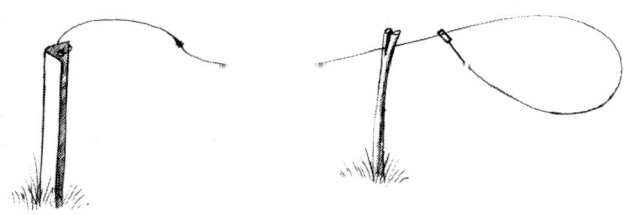

A correctly set fox snare

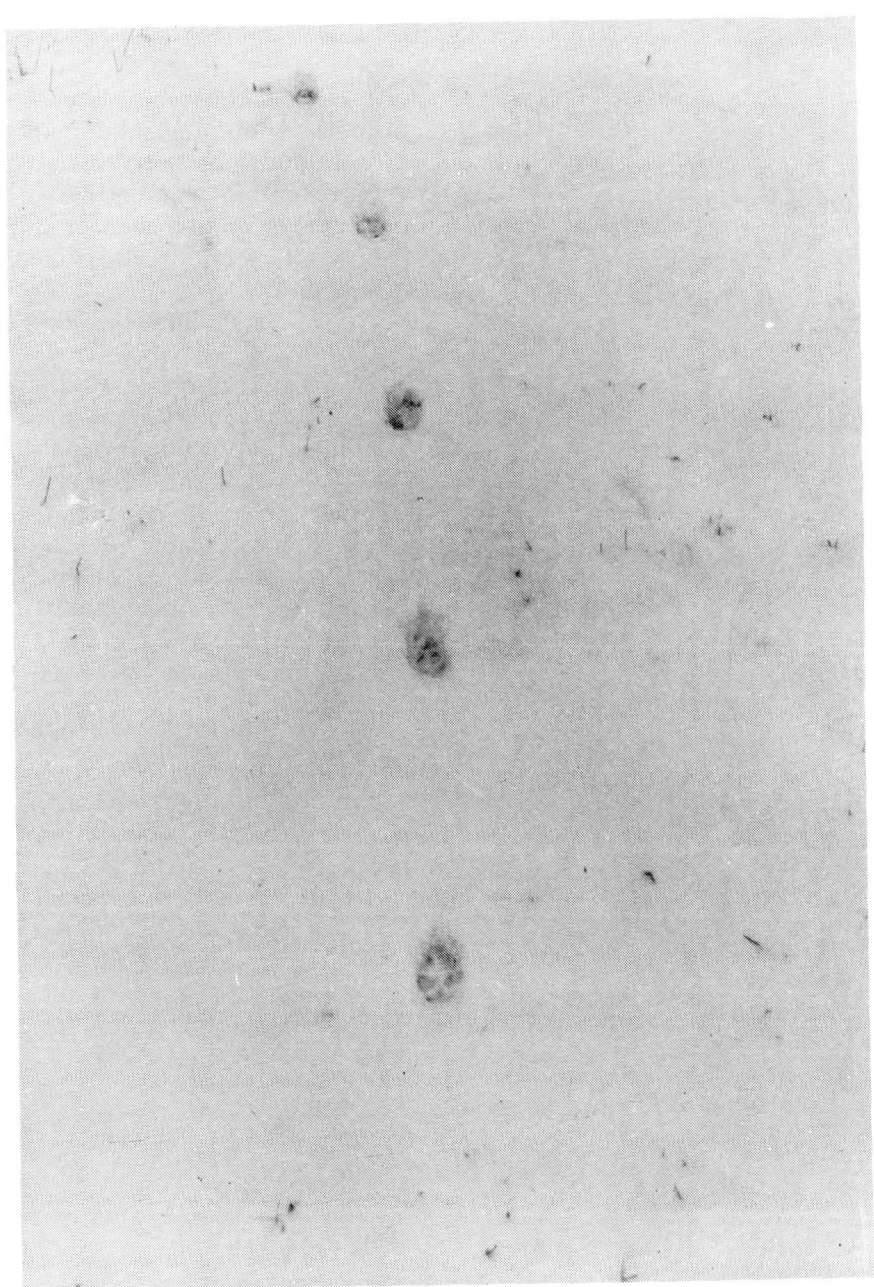
Fox tracks in the snow.

If the situation calls (windy) another pricker might be needed to steady the pear-shaped noose. Sometimes a little camouflage is necessary, particularly around the metal peg, which must be well driven into the ground. Try to make as little mess or disturbance as possible around the snaring point. The snare can be attached to a good fencing post if convenient, or to a stout, heavy log.

REMEMBER THAT A SNARE HAS TO BE CHECKED AT LEAST ONCE A DAY BY LAW.

Normally nothing happens for a night or two. Maybe you will find the snare knocked down. This is because it has been set too low or too high. and sometimes a rabbit or hare will knock them over.

If all goes well, you will find the fox ensnared, normally around the neck, sometimes around the body. If the snare has been fastened to a heavy log you will soon discover which way it has been dragged, particularly on a dewy morning.

If the fox is not already dead, speed is the essential now. If you wish to keep the skin entire, shoot the fox through the ear. This is the quickest method of despatching it. A .22 rifle is best, a shotgun makes a mess.

Snares can usually only be used once in order to maintain their free-running properties. It may be necessary to change the location of the next snare or desist from snaring until you are convinced the run is being used again. Foxes will always move into unoccupied territory.

Before you even consider using snares for foxes, you must think over the following check list, and if there is any doubt, do not use snares.

Lingering death from an unattended snare is horrific and inexcusable.

a) what other animals are about - dogs, cats, sheep dogs.
b) do not snare where there are cattle or deer.
c) do not snare anywhere near a public footpath.
d) lift all snares if fox hounds or beagles are meeting locally.
e) all snares **must** be visited at least one a day, first thing in the morning and again in the evening.

If you have a bird sitting, maybe a pheasant or guineafowl, and at risk from the fox, one emergency way of repelling the fox is to urinate around the nest.

If, however, you have caught a badger, and this sometimes unfortunately happens, you have to release it as quickly as possible. This is not always easy as badger teeth and claws are strong and sharp. If caught by the body, take a sack or coat and place over the head of the animal and with tin snips or piano-wire cutters, cut the noose. If caught by the neck, hold the head down with a pitchfork or forked branch, using good strong gloves on both hands, and then cut the noose. Because of the free-running snare and the build of the badger he is unlikely to have been injured and will amble off when freed.

A Badger and tracks (not to scale)

2. **Fenn Trap:** quarry - rats, grey squirrel, stoats, weasels, mink, feral ferrets.

Set Fenn trap

Although there are other spring traps on the market, we have always found the Fenn Mark 4 to be the most easily used and the most successful. All spring traps must by law be set in tunnels and visited on a daily basis.

A tunnel may be constructed in many ways according to the local materials available and is best kept as natural looking as possible. Here you have to think a little like ground vermin which are shy, but inquisitive.

Location. Most ground vermin hate being exposed, therefore crossing a road, gateway, any open space, must be done at high speed. If there is an inviting hole the other side, possibly with a chink of light coming through i.e. your tunnel, then that is a good location. Blind tunnels seem to work best when baited. Hedges, ditches, walls all offer good areas for hunting and travelling for the vermin; a tunnel trap set at the junction of two hedges, a gap in a wall - possibly especially made, a dry water culvert, beside a chicken house, under piles of wood, around the base of a rotten tree, along streams in the corner of a wood, in short, anywhere that offers cover, a link from one place to another.

Before choosing your location, think through the following points:

1. do not trap close to a public footpath or the public gaze. Traps have a habit of disappearing.
2. do not trap in the way of tractors or other agricultural machinery.
3. do not trap where cows or sheep are present.
4. can you visit the location easily and on a daily basis?

A Brown Rat and tracks (not to scale)

A well caught rat in a Fenn trap. Note the tunnel which has been temporarily moved back in order to re-set the trap.

Construction

a) Permanent. The best tunnels are those made of local materials and look as though they have been part of the hedge, ditch or wall for ever. This also helps to disguise them from inquisitive human eyes. After blending with the surroundings the next consideration with a permanent tunnel is ease of access. You need to be able to service the trap without having to demolish the entire construction, and it must be deep and wide enough for the trap to close without obstruction. Permanent tunnels can be made of bricks, drain pipes, stones, slate, straw bales, roofing tiles, flints, pitch pipe, logs, with possibly a covering of thick turf (thin will wither and go brown). The tunnels need to be about 2′ long. The lead-in needs to be as wide as possible, reducing down to the entrance which then needs sticks or pegs to reduce the available aperture. This is to stop small birds, dogs, cats etc. from gaining access. It is important to catch only what you intend catching, i.e. vermin.

Most permanent trap locations take some time to construct well, but they can be used for many years if in a good location. The floor of the tunnel is best kept natural to the location and where the trap goes make a small indentation, the size of the trap when set and deep enough so the plate is level with the ground. The trap must be set **across** the path of the quarry to ensure a clean kill and the chain buried and firmly pegged to the ground outside the tunnel or wired to a convenient post. This will prevent the trap and contents being carted away by a hungry fox or cat.

When setting the trap *always* set it with the safety catch on and the last thing you do when leaving the site is to flick the safety catch off with a twig. Regular maintenance around traps is important so they don't get overgrown and to ensure that the funnel leading to the entrance is kept clear. Do not, however, 'garden' around the tunnel.

b) Moveable. This tunnel trap is used where there seems to be a build up of ground vermin, lack of local materials to hand, or in an emergency.

Three pieces of wood, 24″ long, 6½″ wide and 6″ high are nailed together to form a box without a floor (inside measurements). The wood can be any type, tanalised will last the longest. Do not make this too heavy, ½″ thickness is quite enough, especially if you have several to carry. The chain of the trap can be stapled to the wood, avoiding the use of a peg and one less thing to carry, or hooked on to the wood to make servicing easier. This type of tunnel can be used at a moments notice if there is a problem somewhere. It is best to make up several of these and keep them outside, weathering, and away from where dogs and cats can urinate on them. Having chosen your site, make use of whatever local material is at hand e.g. dead grass, to disguise the sharp outline of the tunnel.

A successful catch. The lid taken off a permanent tunnel trap.

Always a good place to trap. Under a pile of old logs.

A tunnel trap made with land drains situated beside the perimeter fence. Note the electric fence on the outside.

One less. A large brown rat caught in a moveable tunnel. Note the peg on the left.

Stoat with tracks. (Not to scale)

Weasel. Tracks smaller than stoat. No black tip on the tail.

Hedgehog and tracks (not to scale)

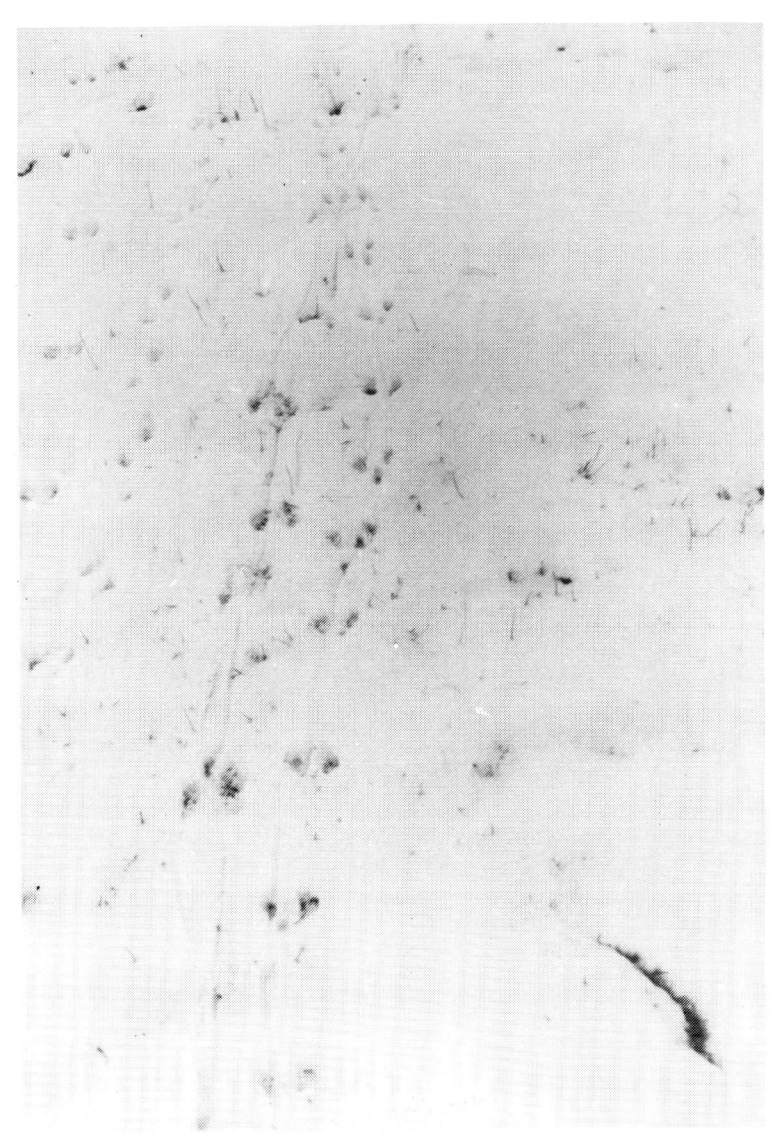

Rat tracks in the snow. Note the tail marks.

How to Handle, Open and Set a Fenn Trap

If right handed, put your left thumb into the loop and pull down, opening the trap, the other hand holding the underneath of the trap. Then bring your right hand over, using the palm of your hand and thumb to hold the trap open and bring the left over in the same way. Put the safety catch on ensuring that the trip lever is hanging loose on the outside of the trap. To set the trap press a little harder with both hands, flip the trip lever over and locate on plate notch so that the jaws of the trap are held open by the lever and not the safety catch BUT leave the safety catch in position until you are ready to leave the trapping place. This is where the importance of gloves comes in - first to disguise your scent and second to protect your hand from the trap should your hand slip.

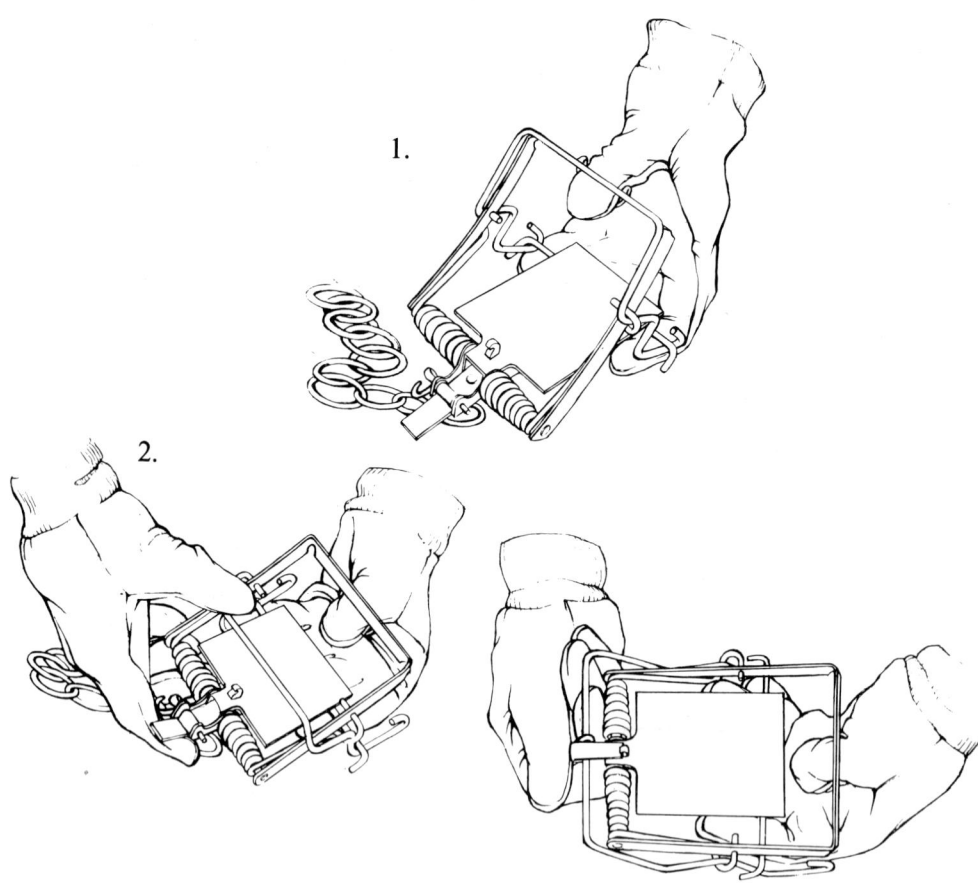

Baits

Baits are not normally necessary on tunnel traps but are used in the following ways:

1) Mink are mad about anything fishy like kipper or tinned cat food. Wire strips of kipper to the plate or put fishy cat food at the blind end of a blind tunnel. Set the trap near water.

2) When a mink, weasel or stoat is caught in a trap, rub the trap all over with the back of the body to give the trap an enticing scent. In the case of a bitch weasel or stoat being caught, empty the contents of its bladder over the trap by gently pushing down the centre of the abdomen towards the hind legs with your thumb.

3) Fresh rabbit liver or entrails placed in a blind tunnel works well for stoats and rats. Ensure the quarry has to pass over the trap to get at the bait.

4) A hen's egg is a good bait in the spring time for rats, placed in a tunnel trap with a trap either side of the egg.

5) The plate of a trap in a tunnel does not normally need to be covered if the tunnel is dark, but it may be necessary to add a little soil, dead grass or leaves if there is much light. Be careful not to impede the working of the trap.

6) A small amount of wheat on the plate of a trap is good for rats. We have caught three at once by this method, but the trap needs to be set fairly hard to prevent mice from springing the trap.

Ferret tracks (not to scale) Mink tracks

A moveable tunnel ready for action. Note the rat run.

Snow reveals all. Note the rat 'motorway' into this permanent tunnel.

Two good places to trap and near to each other. Top, caught red handed under the food bin. And below, a niche by the gate. This trap has caught dozens of rats as they are trying to enter the enclosure.

Tips on Trapping

a) **Rats** always run in the same place for safety and after a while these runs become like paths due to usage or number of rats. Where possible, against a fence, house or barn, put one of the wooden box tunnels over the run. We like to set the trap, but on the safety catch, and allow the rats to run over the trap for a few days. Then set the trap properly. Rats are particularly active after the corn is cut and come into barns, sheds etc. for food and shelter. They start to feed when the sun goes down, so two visits to a well sited trap in the early evening may be necessary. A caught rat may well be eaten by its friends during the night, or other predators, and only the tail may be left by the morning. Always handle a rat with care (this is where your gloves will help) as they carry a disease called Weil's which has been known to be lethal to humans.

b) If a trap has not been correctly set the quarry may not be killed outright. A quick blow with a stick will despatch it, or put the heel of your boot on its head.

c) When trapping for **mink**, you can use the Fenn Mark 6 which is larger than the Fenn Mark 4 and designed for mink and rabbits, but the Mark 4 is well able to cope with mink.

d) Make sure the plate of the trap when operating near water can work without being wedged up by silt or fine soil.

e) Spring a trap that hasn't caught anything once a week to ensure it still works correctly.

f) If your dog accompanies you when trapping make sure it sits well away from the tunnel so its scent doesn't put off the quarry.

g) If you come upon a sprung trap with nothing it it, **mice** may have set off a lightly set trap, or look for hair on the jaws of the trap. Pale grey will be rabbit - a young one small enough to get into the tunnel -or dark brown could be mink. Check that there is enough room for the trap to close properly within the tunnel. If there are feathers in the trap it may be a blackbird or little owl as they love tunnels and are after beetles and worms. Make sure that the entrance is small enough to exclude all but the quarry.

h) Occasionally and unfortunately the odd hedgehog gets trapped. These are difficult to get out, the best way is to open the trap by standing on the base of the trap and pressing lightly.

i) Make good use of fresh snow to discover exactly where vermin are travelling - you will be suprised how many tracks there are - and make notes on the main routes if you haven't already spotted these, for reference when the snow has gone.

3. **Cage Traps:** quarry — mink, grey squirrel

a) **Mink**: Anyone who has seen the devastation a mink wreaks on the wildlife of a lake, stream or river, will understand why these animals should be relentlessly pursued. They are not native to Britain and have no natural predators. One of the first signs of mink is the sudden disappearance of normally numerous moorhens. They also take fish, as they are excellent swimmers, duck, coot, water voles, frogs, leaving a river devoid of its natural inhabitants, let alone the destruction they can cause to domestic poultry or waterfowl. They kill for the sake of it.

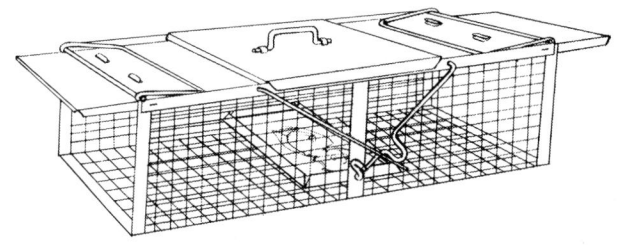

Baited and set mink trap

The cage trap, which should be chained and pegged, particularly when the stream or river is liable to flood, needs to be set near the water's edge, under bridges, in old willow trees, under bank edges, in gulleys and ditches leading into streams. Fortunately mink are fairly easy to catch. The best bait to use is kipper or fishy tinned cat food. They breed readily in the wild with an average litter of five.

The musky smell of mink is strong - a cross between dog and cat. They usually hunt in pairs or trios, so if you catch one in a trap there is likely to be another about too.

The caught mink, by now very hostile, can be killed with an air gun or the cage attached to a cord and dropped into deep water. Once the mink is dead, rub the back end of the mink's body all round the trap to give it a good minky smell. Gassing is the humane method of despatch approved by the authorities.

*This moveable tunnel killed two mink.
The stream is just behind and below the tunnel.*

A cage trap baited with kipper, waiting for the next mink on the river bank. The run comes up from the water on the left.

b) **Grey Squirrel:** although like mink, the grey squirrel is omnivorous, its main destruction concerns trees, particularly young ones and special wood plantations such as hazel, chestnut, walnut, beech and other hardwoods. They will also take eggs and love the food left in hoppers for pheasants.

The cage trap should be placed on the edge of rides through woods, at the base of mature trees, in old roots, anywhere there is evidence of eating, like nut husks or peelings. The trap can be concealed with logs, dead grass and baited with whole maize. The main reason for using a cage trap is in case there are any red squirrels in the area, when tunnel trapping or poisoning is then illegal.

Once the grey squirrel is caught in the trap it can be shot with an air gun or the trap upended carefully into a hessian sack, the trap door opened, and the squirrel can then be despatched using a stick. Scenting the trap as for mink also helps to catch more.

There are other methods of taking grey squirrels (poisoning, page 40) one of which is shooting. If the squirrels nest (called a drey) is on your land then this is in the form of a roundish cluster of twigs sometimes 3' across in the fork of a tree. A drey in use for the current year will have leafy twigs in its construction, but squirrels use old dreys for resting in. The best time to shoot grey squirrels is in the winter months, when the leaves are off the trees, although it is possible to take them in the late summer with a shotgun or .22 rifle while they are noisily feeding. Squirrels use every trick known when being hunted and it is certainly a job for two people. They will freeze against the side of a tree, sometimes high at the very tip, they will creep around the tree as you go round, jump impossible lengths, always having an eye to an escape route, and then in the end disappear in an old woodpecker hole.

It is a good idea to poke out all dreys with a lofting pole (see suppliers, page 48) including old magpie nests, that way new nests will be spotted more easily. **Danger:** lofting poles and electric power lines are lethal.

Grey squirrel and tracks (not to scale)

4. **Mouse traps**: quarry - mice
There are dozens of makes on the market, all with extravagant claims. When a mouse is seen darting between two pieces of furniture or out of a sack, most people freeze with horror. We think it is the speed of travel. Meanwhile, this mouse and the rest of its huge family are chewing their way through food, electric wiring, paper sacks and spreading disease by their droppings and urine.
Two types of traps are described. This does not mean the others do not work, it is just that we have used these two types most successfully.
a) The Little Nipper is excellent in the house, once you have mastered the art of setting it without catching your own fingers, but do keep it out of the way of little hands, cats and dogs. The best bait is not cheese, contrary to most opinion, but chocolate, particularly the holey kind which will stay on the spike. A raisin, a small piece of bacon will do, even a single grain of wheat, the idea being to keep the bait fixed on the spike. If you use these traps in a loft or an area where the trap can be lost (a cavity wall for example) then staple a piece of string to the trap and tie it to something. The smell of a dead mouse in a house is disproportionate to its size and will remain for weeks.
b) The Ketch-All is an American invention and is a galvanised metal box with sliding lid and spring-loaded drum that flicks the mouse into a holding area. This may be used with bait, but will still catch several mice at once without, as the trip is activated by the mouse running through the box. These traps are ideal in aviaries or areas where there are other small animals or birds. To kill the catch, submerge the entire trap in a bucket or tub of water, or use gas.

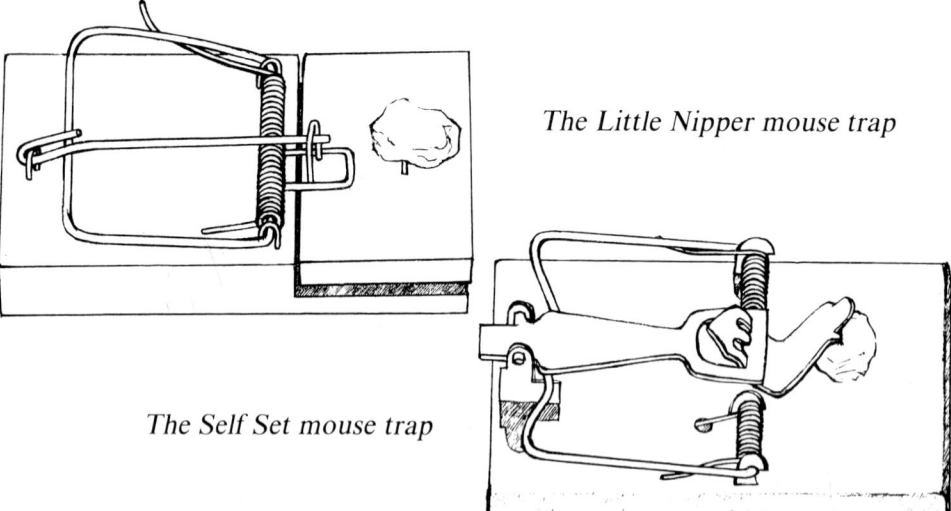

The Little Nipper mouse trap

The Self Set mouse trap

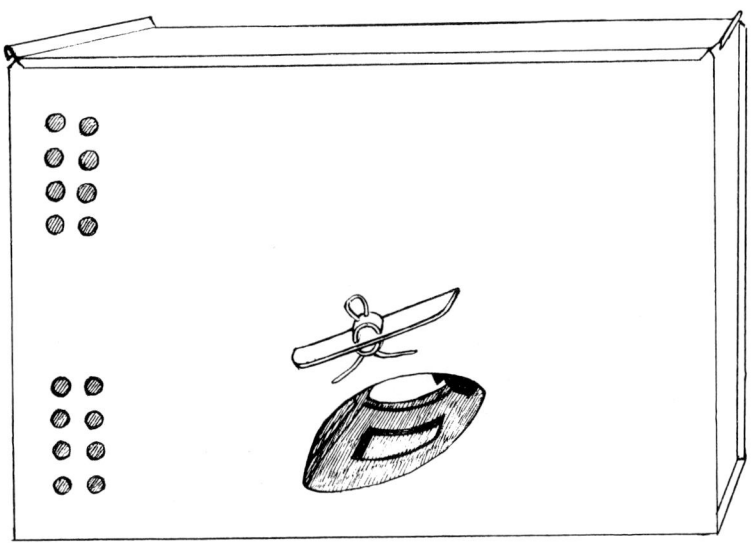

The Ketch-all multi-catch mouse trap.

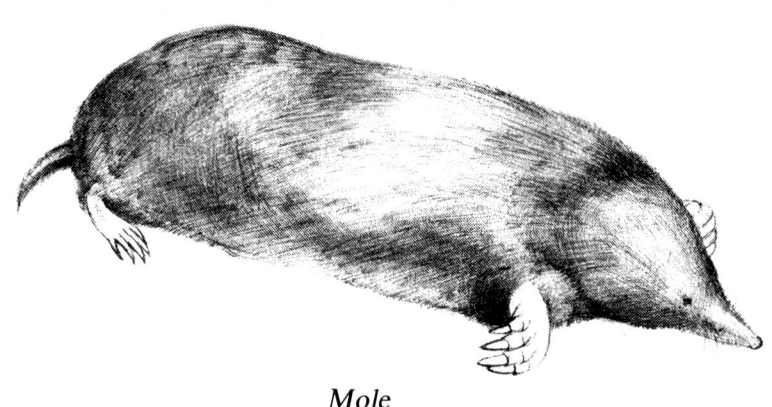

Mole

5. Mole Trap: quarry - moles

Unfortunately the only offence these creatures do is to make a mess. Their holes and runs can be used by weasels.

The most effective way of controlling moles on a small scale is trapping. Moles have poor eyesight, but their noses are well developed. Trapping them is quite an art, again with the use of gloves. They love areas where there is a density of ducks and hens, and in winter near to waterways. We think this is because the manure from the birds attracts worms, and damp soil is easier to work.

In a garden look for a confirmed route between two flower beds, or in a field (clear of livestock) a route between two large mole hills. The grass is slightly raised along the tunnel route, the idea being to trap in as firm ground as possible. With a gardening trowel dig out a small oblong hole, the same size as the mole trap when open, to the depth of the base of the tunnel. If you find a worm in the meantime, spread that on the trip lever of the trap. Place the trap in so that it stands upright and does not rock, set the trap and using the turves that you have dug out, gently bury or surround the trap in order to stop any daylight getting in the hole. We have known these traps to catch within half an hour, but it may take a week to catch a crafty mole.

Moles can be poisoned with Strychnine, but this has to be done by the MAFF Pest Officer.

If you are troubled with moles on a large scale, in a field or parkland, there is a method of shooting them. This is best done in frosty weather, the area to be controlled is rolled with a heavy flat roller behind a tractor, first thing in the morning. Using a shotgun with heavy shot (3's, 'BB's) any new working mole can be seen, approached quietly and shot. The success rate is quite high providing the mole is still pushing up earth when you shoot.

Mole. Set the trap on a main run between two large mole hills. This trap is sprung.

Killed outright.

6. **Bird cage traps**: quarry - rooks, carrion crows, magpies, jays, jackdaws, feral pigeons, collared doves, sparrows, moorhens (only between 1st Sept & 31st Jan.)

These come in two main types and different sizes. See diagrams.
The first is the funnel cage trap, where the entrance is from the ground. It can also have a funnel in the roof. This basically is a wire mesh box with a door for access on one side.
The second type has an entrance from the top through a form of a ladder. Again, this has a door for access on the side.
With the exception of the small sparrow trap these bulky traps need to be made in bolt-together sections. This way they can be carried on a trailer from site to site.
Location is all important. These traps are used when there is a crowd of birds thieving eggs, cattle food, crops etc. Place the trap near to the area of pilferance, beside a hedge, wall, under a tree, preferably dead so they can use the branches as perch points, near straw or hay stacks, amongst poultry houses.
The next thing is to bait the trap with one side taken off and baited for a week so that the birds become dependant on the food and forget about what they were pilfering before. The bait, which must be put in undercover of darkness every night can range from offal to white bread, maize, eggs. It is tempting to peek at a cage trap during the day, but the area wants to be kept as quiet as possible with other domestic animals and birds temporarily fenced off. After a week, replace the side that was open, so access is now only via the ladder or funnels. Bait again in the evening when dark and keep away from the trap until it is dark the following evening. This way the other quarry birds will not become suspicious or shy of the trap. Block up the funnels and ladder with sacks or coats and enter the trap with a torch. Release any non-quarry birds which have got into the trap and then catch and kill the quarry birds cleanly with a stick across the back of the neck and take them away. Try to leave no trace of your entrance and put in fresh bait. Do the same thing the following night, when most of the quarry birds should have been caught. Rarely does a cage trap work in the same location more than two nights running. It is best to move it around to different places or not to use it at all for a month or so. These birds are wily and intelligent and once frightened will not go near a cage trap again.

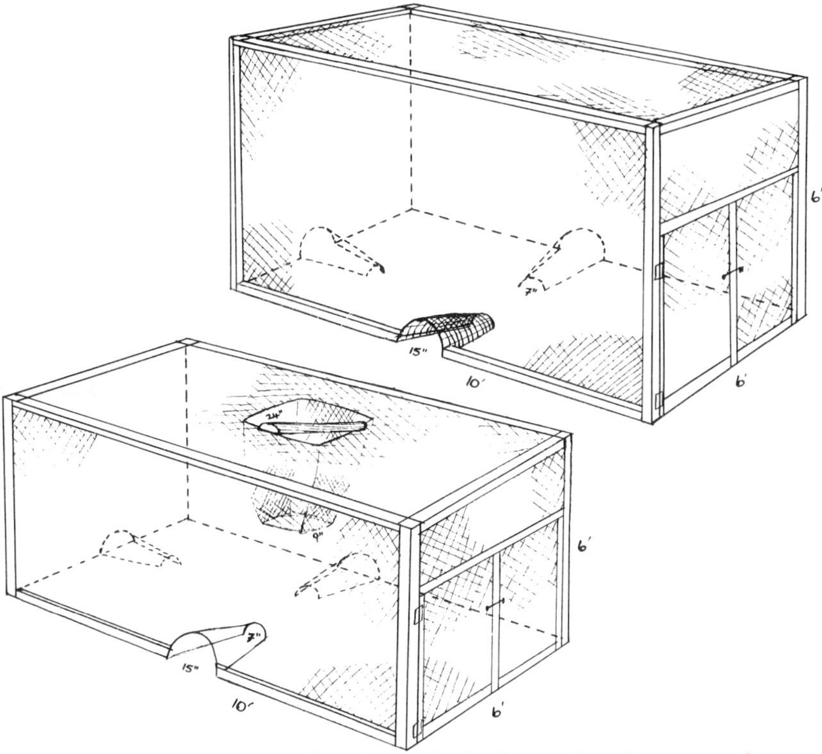

These cage traps are sectional and bolted together for ease of transport. White bread is a good bait.

Jackdaw

Carrion Crow and track.

Rook - Note differences of beak, shape, trousers and general untidiness between rook and carrion crow.

Magpie and track

Jay

Moorhen - *vicious towards ducklings, and will eat eggs.*

7. **Poison (bait boxes):** quarry - rats, mice, grey squirrels
We dislike the indiscriminate use of poison because of the danger to other animals and birds for whom it was not intended. The systems of using these poisons can work well, providing every care is taken. It is sometimes necessary to use poisons when rats become trap shy, or when the use of traps is hazardous.

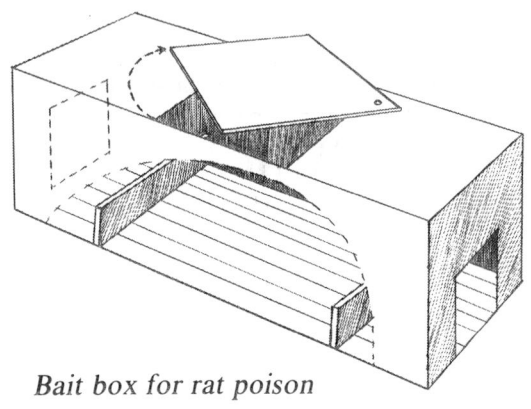

Bait box for rat poison

Rats, especially in winter, are a menace under or around barns, sheds, aviaries, pigsties, hen houses etc. and can climb vertical walls with ease. They are not only destructive to buildings, making holes and burrows, but eat and contaminate feed. They are carriers of various diseases which can be passed on to livestock. It is not enough merely to try and bar them from an area. We remember seeing a new collection of ornamental pheasant aviaries which were quickly colonised by rats. The floor of the aviaries were then covered in ½" wire mesh and the rats merely ran about underneath the mesh, in broad daylight, eating their way in via the wood surround. Simple bait boxes would have cleared up this problem, a problem which aviarist, duck, poultry or pheasant keeper knows only too well.

As rat poisons are based on various anti-coagulants, the poison has to be eaten over three or four days at least to take effect. This is where bait boxes are ideal. The first design is for use where there is no risk of any other animals or birds entering the box - warehouses, food storage areas, barns.

The second is our own design and invention which we use at The Domestic Fowl Trust. It is a dual purpose box, being used either as a tunnel with a Fenn trap or as a poison bait box. Either way, the trap or poison are completely safe from wild birds, domestic fowl, pets etc. so this can be used inside or out, and around the garden. If used as a poison box, lace the poison with icing sugar which rats find irresistable. Remember to top up the poison regularly and also to change the brand of poison used every couple of months so the rats don't become immune to one chemical. Weathered boxes will work quicker than new ones as they will have less human smell on them.

If used as a tunnel, the DFT Ratbox is best placed along a known run. Put a half inch layer of wheat in the centre and the trap set on top of the wheat. This trap kills every time as there is only place the rat can land - in the centre of the plate.

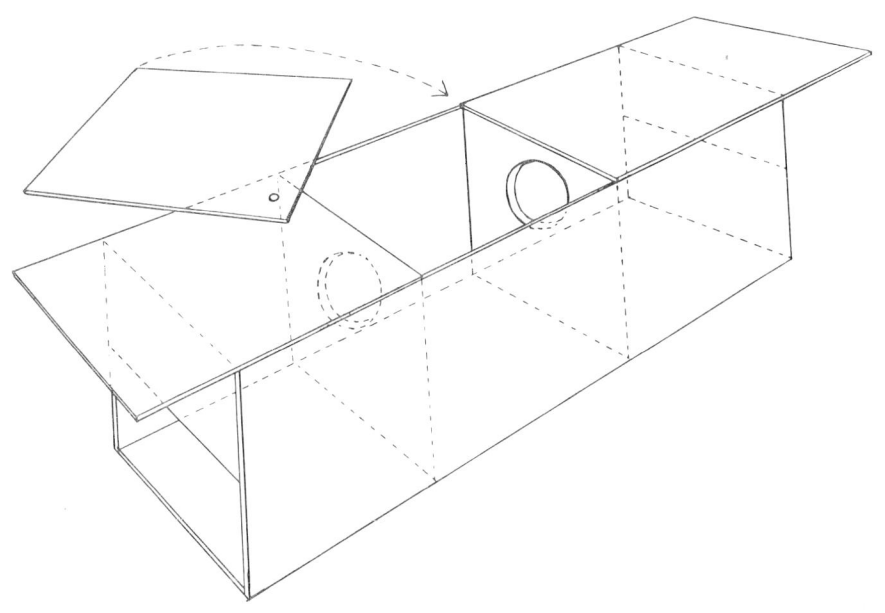

DFT Ratbox

Mice will eat rat poison, but they become shy of poison quicker than rats, and unlike rats will die in the open when other animals or birds might eat them.

Grey Squirrels can also become trap or cage shy and poison can be used, especially where daily visits are impracticable. There are certain areas of the UK where poisoning of grey squirrels is illegal due to the presence of red squirrels: All of Scotland, Northumberland, Cumbria, Durham, Lancashire, East Suffolk, Norfolk, Isle of Wight, Anglesey, Gwynedd, Clwyd, part of Powys (Montgomery), part of Dyfed (Cardigan and Carmarthen).

Preparation of poison bait: Only warfarin poison may be used by law. Using gloves, put 12 Kg or 26 lbs of wheat into a dustbin or plastic bag (fertiliser type), add 250 ml of concentrate liquid warfarin and mix thoroughly. Then add a further 250 ml of concentrate and mix thoroughly again to ensure that all the grain is coloured red evenly. Leave this to soak for three hours. Do not exceed the prescribed dosage.

Siting: The siting of the hopper is all important, not only to have maximum effect but also to ensure there is no danger to other wildlife in the area such as deer, badgers, pheasants etc. The hoppers can be placed in old tree stumps, wired to trees or posts, usually one per acre of woodland is enough placed 100 yards apart in a grid formation. Ensure that the hopper is fixed and on the level. This stops any spillage and any rain water from running up the tube and solidifying the poison. The best locations are in woodland with a clear floor, i.e. leaves or pine needles and not brambles. If the woodland is open to the public do **not** use this form of control. The public are very sensitive to the control of grey squirrels, confusing them with the red, and vandalism and loss of equipment will ensue.

Baiting: once the hoppers are in place leave some trails of whole maize leading to the hopper. Squirrels love maize and the bright yellow kernels can easily be spotted. Pre-baiting is essential for about a week in order that the squirrels become semi-dependent on the maize for their daily diet, taking a mixture of wheat and maize from the hopper itself. When the hoppers are nearly empty, introduce the poison bait. It will normally take about a week for the hoppers to empty, depending on the numbers of squirrels in the area, so regular visits are essential. Warfarin poison takes between 7 - 10 days to kill a squirrel, depending on the quantities eaten. There is no smell to warfarin and the squirrels appear to die of old age. Pick up any dead squirrels and leave the hopper in place for about a month. When checking the hopper make sure the poison is flowing properly and that there are no obstructions in the tube. Move the hoppers on with the squirrels into new areas, but it may be adviseable to poison again in the same area the same year.

Important: It is vital to mark clearly with a poison sticker all the hoppers so that if there are any problems, all parties are aware of the contents.

A suitably sited grey squirrel poison hopper

Beware of using poison near poultry or pheasants. If the poison is eaten, particularly the anti-coagulants, this causes bleeding in the lungs with the bird throwing up blood. The normal antidote of Vitamin K injections does not seem to work well on birds, and death usually follows in a few days.

Do not use a shotgun around ponds or streams where duck or geese are kept. The lead in the shot can be ingested by the birds and slowly poison them.

You will note that we have not included gassing in this book. This is not for the amateur and in the wrong hands can be lethal to unintended victims.

8. Ultrasonic Devices: quarry - rats, mice, feral pigeons, starlings, moles.

In areas where more conventional means of controlling vermin and pests is sensitive or impossible, there are several ultrasonic systems on the market which can be used successfully. Ultrasonic sounds work well providing the target area is completely covered by them.

Rats, mice - The frequency in which rodents communicate is 18-30KHz and some devices go as high as 100 KHz. The pattern of sound and random pulsing continually changing disorientates the rodents, upsetting the nervous system, the appetite and the desire to mate. They either die or are driven out of the area.

Children can hear up to 20 KHz and adults up to 15 KHz, but these frequencies and sounds do not disturb dogs, cats, farm livestock, televisions or burglar alarms. These ultrasonic devices can be used in warehouses, farm buildings, houses, mills, factories, grain stores, foodstuff areas and are either mains or battery operated.

Installation should be in an enclosed building on the floor or a low shelf. Remember that ultrasonics cannot go through solid objects, but bounce off them, so ensure that there are enough speakers to saturate the infested area with ultrasonic beams. Certain materials are more reflective than others like glass, metal, brick walls, plyboard, and others more absorbent like hessian sacks, straw bales. Try to stack foodstuffs away from walls and on pallets.

If you are in doubt about the degree of infestation in a building, sprinkle some flour or powder on the floor and this will highlight the numbers and direction of the rodents.

Feral pigeons, starlings The ultrasonic frequencies for birds are lower than for rodents and can be heard by humans. Some devices are applicable to both birds and rodents at the flick of a switch. Because ultrasonic sound dissipates quickly out of doors, it is necessary to have several speakers for effective control. Again, bouncing the beams off buildings gives greater coverage. Time switches help as these birds do not feed at night. A polypropylene line, developed for horticulture, is useful when stretched tightly between stakes as every breath of wind produces a different hum frequency, changing between subsonic, sonic and ultrasonic.

Moles There is a German device available using a vibrating conductor. Waterproof and run on 1.5 volt batteries (lasting 2-3 months) the mole repeller is pushed into the ground where moles are working and switched on. It works 24 hours a day and produces 500 piston-like up and down motions per second. The area covered by this thumping sound is approximately 1000 square yards.

Identifying vermin from carcases

All vermin have their own special methods of killing and eating, so this makes it fairly easy to identify which vermin is responsible for a carcase.

Fox: head bitten off, birds left maimed, bitten across the back, sometimes buried, birds chased over a wide area, large clumps of feathers, most of carcases in the house or pen, signs of entry such as wire pulled out, wood bitten through.

Rat: teeth marks around the top of the neck, carcase stripped clean down to the skeleton, eggs broken and licked clean.

Crows: flesh stripped on the neck. Will take goslings and ducklings up to 6 weeks of age. Holes pecked in eggs or eggs carried away.

Rooks: rarely attack adult birds, great egg thieves.

Mice: will nibble carcases, usually starting at the tail end.

Weasel and Stoat: teeth marks around top of neck. A stoat will kill several birds in one night.

Mink: small bite around base of head and neck, several birds dead at once, no sign of entry, heads off smaller birds.

Magpie: young ducklings or chicks totally disappearing, eggs broken or carried away.

Feral cat: bitten across the back.

Brown owl: total disappearance of chicks with no sign of a struggle.

It is always worthwhile keeping a record of vermin caught. This can help in future years to be more alert about the dangers of vermin and their most active seasons.

Useful by-products of vermin

Skins and feathers

Having trapped or taken your quarry, it seems sensible to make something either in cash or kind from the pelt or feathers. There are several companies that will buy these from you, particularly if the skins are fresh and well skinned and the wings presented correctly.

1. Fresh skins from: weasels, stoat, mink, fox

2. Dried wings from: jays, magpies, rooks, crows.

3. Dried tails from grey squirrels.

Prices will vary, particularly for fox and stoat pelts as they are at their best in a hard winter.

1. Skin the animal as soon as possible. Start under the chin and slit carefully with a very sharp knife down to the tail. Do not pierce the paunch. It may be necessary to skin out the head, the purchaser will advise. Cut across the inside of the forelegs down to and around the pads. Cut across the inside of the hams to the hock joint and around it. Cut round the genitals and pull the bone from out of the tail or brush by using a spliced hazel or ash stick to hold the skin while you pull out the bone. It gets easier with practice.
The purchaser will usually then want the skin rolled up, fur side out to keep it moist and sent off by the quickest method, but will advise.

2. Cut of the wing with a strong pair of scissors as close to the body as possible. Spread the wing out into the flight position and put near a source of heat with a weight on top. About a week for the smaller wings should suffice, longer for the larger ones. The flesh should be completely dried out, hard and odourless. Magpie tails can also be dried in this way and wings and tails sold to the feather dealers for fly tying.

3. Cut off the tail where it joins the body, stroke the hair until it lies at right angles to the bone and then dry near a source of heat until the flesh is dry and odourless. Also used for fly tying.

How to air dry and cure a small skin for your own use from trapped vermin

Having skinned the animal, scrape all surplus fat, meat, blood and veins from the skin with a blunt knife. Wash the skin in a biological detergent to get any stains or blood off. Rinse immediately in warm water and add some fabric conditioner. Pat surplus water off with towels or absorbent paper, do not wring. Stretch the skin out on a board or door, fur side down, and nail using brass or non-ferrous pins or tacks. Try not to overstretch the skin and nail around the outside about 1" apart.

Curing: Paint with a paste of 3 tablespoons of bicarbonate of soda to ½ pint of paraffin. Allow to wind dry, avoiding exposure to hot sun. Small pelts will require two coats, larger pelts such as fox, four coats of the paste, allowing each to dry. the pelt will look like parchment and be slightly powdery. Remove from the board, taking out the nails. Small skins should take 1-3 days and 3-5 days for larger skins.

Finishing: firmly wire brush the skin to take off any bits of flesh, taking care not to perforate it. Then gently rub in some olive oil and egg yolk or flour in equal quantities and leave for 24 hours. The skin now needs to be 'worked' to soften it, either by rubbing the surfaces together or rolling between finger and thumb or on larger skins pulled to and fro over the back of a chair or trestle. Start at the outside and work inwards. Elbow grease and patience are the main ingredients. Sprinkle and rub in talcum powder and trim off the hard outsides edges of the skin with a knife, **not** scissors.

The skin is now ready for use. To check if you have cured the skin properly, try to pull out a tuft or two of hair - it should be firmly anchored.

If you require a bird or animal to be stuffed, wrap it fresh in absorbent paper, taking care that the feathers or fur are lying the right way and as little blood as possible present. Then wrap in a plastic bag and seal and deep freeze. This way the subject can be kept for a long period ready for when the taxidermist can take it.

NB. No registered taxidermist will accept any wildlife that is not vermin if it has been shot or taken illegally.

Brown Owls, Badgers, Hedgehogs, Feral Cats

All the above species are protectd by law, but there comes a time when one of the above becomes a pest, the law making provision for this and allowing the use of a shotgun, providing that it can be shown in a court of law that the species is causing serious damage.

Brown Owl.

Little Owl — may occasionally stray into a tunnel trap, but is a protected species.

Brown Owls are normally harmless, but can get a liking for young pheasant poults. We have had them take young chicks out of a stable at night. The owl was seen to enter by the ventilation window which was afterwards wired off, but when you have a tawny attacking poults in a release pen the only way may be to shoot it. The trouble with such a rogue bird is that it does not just kill one, but dozens at a time.

Badgers rarely are any trouble, but there may be an occasional rogue one which will take poultry, pheasants and lambs. A rogue badger has normally been thrown out of the set by the rest of the family because he is too old, or injured. He becomes a tramp, living rough above ground, smelling strongly and becoming a scavenger. This must be dealt with quickly under the supervision of the MAFF Pest Officer. There may be times when the local population of badgers explodes and becomes too much for the food in the area to support. Before the badgers' protection, these explosions would have been dealt with by the local gamekeeper, but you must now call in the local MAFF Pest Officer.

Hedgehogs normally cause little damage and it always nice to see them around, keeping the slug population under control. They can get a liking for eggs and become terrible thieves. Occasionally there are odd animals who go further - we found the remains of a broody hen, having been eaten from the back to the front by a hedgehog which had got in under the back of the coop and was found curled up in the carcase. If you have a proven troublesome hedgehog it may be legally killed with a gun, or better still take it for a long ride in the car.

Feral Cats. Of all the predators in the countryside the feral cat is the most difficult to deal with. A feral cat is a domestic cat that is living permanently wild in the countryside. This sort of cat has all the cunning of a fox, and more, and although sometimes only killing for its immediate needs, can wreak havoc among pigeons, birds in aviaries, pheasants, poultry and duck. A shotgun or rifle is the only method of taking this nuisance. If this method is not available to you, try and entice the feral cat into a shed or stable then call in the local RSPCA officer who will have all the catching equipment and can take the feral cat away.

Feral cat track.   *Domestic dog track.*

Herons and cormorants are protected under the Wildlife and Countryside Act 1981, but an occasional licence may be issued by the MAFF only if they can be convinced that all other methods of prevention have failed, and where someone's livelihood is at stake.

Suppliers and useful addresses

A. Fenn, Hoopers Lane, Astwood Bank, Redditch, Worcs. Tel: 052789 2881. *Fenn traps, snares.*

Quadtag Ltd., Corrys, Roestock Lane, Colney Heath, Herts. AL4 0QW. Tel: 0727 22614. *Traps, snares, cage traps.*

Alton Metal Works Co. Ltd., Mill Lane, Alton, Hants. *Grey squirrel poison hopper.*

Fuller Engineering Ltd., 83 Offington Avenue, Worthing, W. Sussex BN14 9PR. *Grey squirrel poison hopper.*

Alpe Thermo Products, 24 Willsbridge Hill, Willsbridge, Bristol BS15 6EY. Tel: 0272 324411. *Grey squirrel poison hopper.*

Croptex Horticulture Ltd., 149 High Street, Ongar, Essex CM5 9JD. Tel: 0277 363173. *Hummline bird scarer.*

E. Parsons & Sons Ltd., Blackfriars Road, Nailsea, Bristol. Tel: 0272 854911. *Lofting poles.*

E. Veniard Ltd., 138 Northwood Road, Thornton Heath, Surrey CR4 8YG. Tel: 01 653 3565. *Buyers of wings, tails and skins.*

Horace Friend Ltd., Nene Quay, Wisbech, Cambs. Tel: 0945 582947. *Buyers of skins.*

The British Association for Shooting and Conservation, Marford Mill, Rossett, Clwyd, LL12 0HI. Tel:0244 570881

The British Field Sports Society, 59 Kennington Road, London SE1 7PZ. Tel:01 928 4742.

The Country Landowners Association, 16 Belgrave Street, London SW1X 8PQ. Tel: 01 235 0511.

The Game Conservancy, Fordingbridge, Hampshire. Tel: 0425 52381.

Ministry of Agriculture Fisheries and Food, Agriculture and Development Advisory Service, Whitehall Place, London SW1 2HH.

The Forestry Commision, Flowers Hill, Brislington, Bristol BS4 5SY. Tel: 0272 713471.

Farming and Wildlife Advisory Group, The Lodge, Sandy, Beds. Tel: 0767 80551.

Nature Conservancy Council, North Minster House, Peterborough PE1 1UA. Tel: 0733 40345.

The World Pheasant Association, P. O. Box 5, Church Farm, Lower Basildon, Goring, Reading, Berks. RG8 9PF. Tel: 0491 671271.

The British Waterfowl Association, c/o Mrs. R. Taylor, Gill Cottage, New Gill, Bishopdale, Leyburn, N. Yorks. DL8 3TQ. Tel: 0969 663693.

The Wildfowl Trust, Slimbridge, Gloucestershire GL2 7BT. Tel: 045389 333.

Shooting Times & Country Magazine, 10 Sheet Street, Windsor, Berks. SL4 1BG. Tel: 07535 56061.

Countryside Monthly, Corrys, Roestock Lane, Colney Heath, Herts. AL4 0GW. Tel: 0727 22614.

The Field, The Harmsworth Press Ltd., Carmelite House, Carmelite Street, London EC4Y 0JA. Tel: 071 353 6000.

Domestic Fowl Trust, Honeybourne, Nr. Evesham, Worcs. WR11 5QJ. Tel: 0386 833083. *Traps, snares, DTF Ratbox, Larsen traps.*

If you have any difficulty obtaining any of the products mentioned in this book, please contact the Domestic Fowl Trust.

INDEX

bait 23
badger 5,9,13,46

carrion crow 5,34,36,43,44
cat, feral 5,43,46,47
collared dove 34
coot 5
coypu 5
crow, hooded 5

fencing 7,8
ferret, feral 5,14-26
fox 5,7,9-12,44

gull, greater black-backed   5

hedgehog 5,20,26,46,47

jackdaw 5,34,35
jay 5,34,37,44

Law 6,12,47
lofting pole 29

magpie 5,34,37,43,44
mice 5,7,26,30,31,38-41,42,43
mink 5,7,14-28,43,44
mole 5,31-33,42
moorhen 5,27,34,37

otter 6
owl, brown 43,46
owl, little 26,46

poison 38-41
polecat 5

rabbit 5
rat 5,7,14-26,38-41,42,43
rook 5,34,36,43,44

skins 44,45
snare 9-12
sparrow 5,7,34
squirrel, grey 5,14-26,29,38-41,44
squirrel, red 40
starling 34,42
stoat 5,7,14-26,43,44

traps, cage 27-29,34,35
traps, mouse 30
traps, tunnel 14-26
ultrasonics 42

weasel 5,14-26,43,44
woodpigeon 5

NEW DEVELOPMENTS

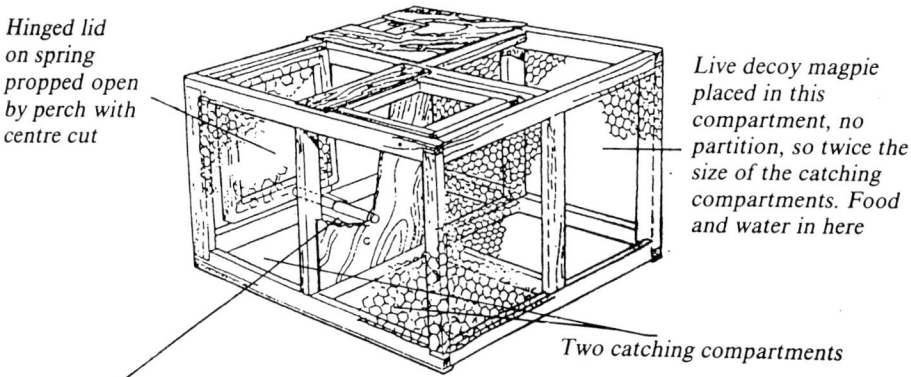

Hinged lid on spring propped open by perch with centre cut

Live decoy magpie placed in this compartment, no partition, so twice the size of the catching compartments. Food and water in here

Two catching compartments

Magpie, trying to join decoy stands on perch which drops under its weight, thus closing the lid

Larsen Trap

Larsen traps were designed by a Danish gamekeeper in the 1950's. In Denmark, it has been suggested that this trap alone was responsible for a significant reduction in the national magpie population from 1965.

Larsen traps will catch all corvid pest species (i.e. crows, magpies, jackdaws, jays, rooks) at all times of year, but their particular value is in catching crows and magpies when these birds set up their breeding territories. The trap mechanism involves a spring door to each catching compartment, which when set is held open by a split perch. In order to enter the trap, birds the size of a magpie or crow inevitably drop onto the perch. The perch gives way, and the bird's momentum takes it past the bottom of the door, which flips up - et voila! Because Larsen traps are small they can easily be moved around. Traps can be moved to deal with specific pairs of crows or magpies, and a few traps can thus be made to cover quite a large area.

Larsens are live-catch traps. Why catch alive? Because of the risk of catching birds other than corvids. Virtually all such non-target birds are protected by law, and must be released alive and unharmed. Having said this, we have experienced very few captures of non-target species in Larsen traps, another point in their favour. Finches and tits often visit them, but are too small to trigger the mechanism, and can in any case escape through the mesh sides. Of course, many legally protected bird species suffer from corvid predation on their eggs or young, and the Larsen trap is potentially an effective tool in the conservation of these birds too. There is no 'natural balance' between corvids and the birds they prey on, because they also feed to a great extent on other foods provided -directly or indirectly - by man.

A second reason to catch corvids alive is that each may in turn be used as a call-bird to attract further captures. In this way, given that you operate several traps, the whole effort quickly grows to an effective scale within a single breeding season.

Operating with call-birds - A call-bird is a previously-caught magpie or crow, which is kept alive in the special compartment of the trap. Uncaught territory holders think a single call-bird is an intruder, and will try to drive it away. They are very aggresive, and if the trap is left in peace, few are so shy that they will not get caught. In a scientifically conducted experiment by the Game Conservancy in 1989, traps with call-birds were fifteen times more efficient at catching crows, ten times more efficient for magpies.

Taking care of your call-birds - Look after your call-birds. They will work best for you when in good health, because then they move about more in the trap and catch the attention of territory holders. They also call vocally, but won't do this if miserable. Visit each call-bird once a day to renew food and water. If they are seen by territory holders to be actively feeding, they will arouse special jealousy. (Apart from this, you are bound by animal welfare laws.) These birds drink a lot of water - earthenware hamster bowls make very suitable non-tipping receptacles. Various kinds of food are suitable, but we have found 'sausages' of brawn-type dog food, fed with bread, to be very convenient. If carrion is fed, make sure it is cut up or at least cut open. Make sure the call-bird has a perch, and shelter from hot sun, rain or wind.

After a while call-birds get tame and phlegmatic. Recently caught territory-holders make the best call-birds, as they are more restless and aggressive, so simply move new captures to act as call-bird in a different territory, and keep a turn over of call-birds.